Activities for
BEGINNING AND INTERMEDIATE ALGEBRA

Student Edition

Debbie Garrison
Judy Jones
Jolene Rhodes
Valencia Community College

Brooks/Cole Publishing Company
I(T)P® An International Thomson Publishing Company

Pacific Grove • Albany • Belmont • Bonn • Boston • Cincinnati • Detroit • Johannesburg • London
Madrid • Melbourne • Mexico City • New York • Paris • Singapore • Tokyo • Toronto • Washington

Assistant Editor: *Linda Row*
Marketing Representatives: *Maureen Riopelle, Alston Mabry*
Editorial Assistant: *Melissa Duge*
Production: *Dorothy Bell*
Printing and Binding: *West Publishing Company*

For more information, contact:

BROOKS/COLE PUBLISHING COMPANY
511 Forest Lodge Rd.
Pacific Grove, CA 93950
USA

International Thomson Editores
Seneca 53
Col. Polanco
11560 México, D. F., México

International Thomson Publishing Europe
Berkshire House 168-173
High Holborn
London WC1V 7AA
England

International Thomson Publishing GmbH
Königswinterer Strasse 418
53227 Bonn
Germany

Thomas Nelson Australia
102 Dodds Street
South Melbourne, 3205
Victoria, Australia

International Thomson Publishing Asia
221 Henderson Road
#05-10 Henderson Building
Singapore 0315

Nelson Canada
1120 Birchmount Road
Scarborough, Ontario
Canada M1K 5G4

International Thomson Publishing Japan
Hirakawacho Kyowa Building, 3F
2-2-1 Hirakawacho
Chiyoda-ku, Tokyo 102
Japan

Printed in the United States of America

10 9 8 7 6 5 4 3

ISBN 0-534-35355-X

This book is dedicated to

Our Families,

Our Colleagues,

and

Our Students,

Without whom none of this would have been possible

With Thanks to our editor and the staff of Brooks/Cole Publishing.

DG, JMJ, JR

TO THE STUDENT

Algebra is not a spectator sport. It is important for you to become actively involved in your learning.

These activities will provide you with opportunities for discovering algebra concepts, reviewing mathematics topics, working with other members of your class as a team, using oral and written communication, and relating algebra to other disciplines.

Your instructor will let you know which activity you will be working on and may provide additional materials or instruction during the activity time. These activities do not have to be done in the order presented in the workbook.

We have used these activities with our students. They enjoy the chance to be more involved in their learning. We hope you do too.

Debbie Garrison Judy Jones Jolene Rhodes

TABLE OF CONTENTS

Applying Formulas

CROSS - NUMBER PUZZLE

This activity provides practice with basic operations involving signed - numbers. The rules for signed - number arithmetic follow. Review the rules then complete the cross - number puzzle using the clues provided. Each block may contain only one digit. For all negative answers, the first block will contain a negative (opposite) sign and the first digit of the answer. The first clue has been completed for you.

Rules for signed - number operations:

Multiplication and Division

- When multiplying (or dividing) two numbers with the same sign, the product (quotient) is positive.
- When multiplying (or dividing) two numbers with different signs, the product (quotient) is negative.

Addition

- When adding two numbers with the same sign, the sum has the same sign as the numbers being added.
- When adding two numbers with different signs, subtract the absolute values of the two numbers and give the sum the same sign as the larger absolute number.

Subtraction

- When subtracting two numbers, add the opposite of the minuend and follow the rules for addition.

Rules of Order:

When an expression contains more than one arithmetic operation:

- First, simplify any operations inside parentheses or other grouping symbols such as, absolute value signs, radicals, or the numerator or denominators of fractions.
- Second, simplify any exponents in the expression.
- Next, perform any multiplications or divisions. If there are more than one, perform the operations as they occur from left to right.
- Last, perform any additions or subtractions. If there are more than one, perform the operations as they occur from left to right.

CLUES

ACROSS

1. $-438 - (-115)$
4. $35.15 \div .37$
6. $(4)(5)(-17)$
9. $50 - (-7)$
10. $8 \times |-6 - 1|$
11. $17 - (-11)$
12. $-23 + (-8)(6)$
14. $-32 \times (-3)$
16. $5 \times 2^3 + 3$
18. $-36 \div -3$
20. the opposite of (2×71)
22. $(-97 \times 54) + (-210)$
25. $(7)(151)(3)^3$
27. $-919 \times -8 \div 2$
29. $-7 + 9 \times 6$
30. $-13 + 28$
32. -33×-3
34. $70 - (-2)$
36. $-5^2 \times 31$
37. $(-13)(2)(-17)$
39. $2^3 \times 3^4$
40. even integer between -80 and -84
41. $16 \times 40 - 215$

DOWN

2. $\frac{1}{3} \times 753$
3. odd integer following 35
4. $-570 \times \frac{-1}{6}$
5. $470 - -99$
6. -2^5
7. $-36 + 104 \times 5$
12. $25 \times -3 - 4$
13. $(300 + 47) \times 3^2$
15. $9(700 + 27)$
17. $6 \times 7 - 3 \times 4$
20. -11×17
21. $5 \times 11 - 10$
23. $(-2)(-23)$
24. $9 \times 100 + (-21)$
25. $2^2 \times 61$
26. $(130 - -1) \times 7$
28. $170 \times 5 - 158$
31. $13(4)$
33. $2(-80 + -9)$
35. $2^3 \times 3 \times -31$
36. -2×37
38. $2 + 40 + 4 \times -5$

CROSS - NUMBER PUZZLE

1 -3	2 2	3 3	■	4	5	■	6	7	8
■	9		■	10		■	11		■
12		■	13	■	14	15	■	16	17
	■	■	18	19	■		■	■	
■	20	21		■	22		23	24	■
25				26	■	27			28
29		■	■	30	31	■	■	32	
	■	33	■	34		■	35	■	
■	36			■	■	37		38	■
39			■	40		■	41		

WORKSPACE

SIGNED - NUMBER MAGIC

This activity provides practice with adding signed numbers (including rational numbers).

1. Complete the following magic square using integers such that $-9 \leq x \leq -1$. Each row, column and diagonal must add up to the same value.

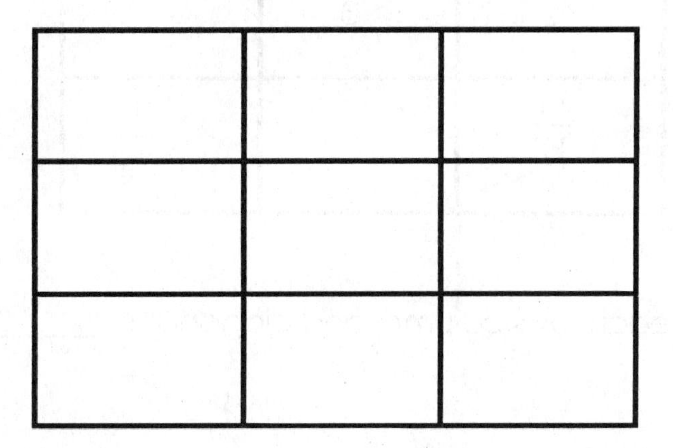

The sum of each row, column, and diagonal is _______________.

2. Complete the following magic square using the integers such that $-1 \leq x \leq -16$. Each row, column and diagonal must add to the same value.

-16		-3	
	-11		-8
-9	-7		
		-15	-1

The sum of each row, column, and diagonal is _______________.

3. Complete the following magic square. Each row, column and diagonal must add up to the same value. All non-integer answers must be written as fractions in reduced form.

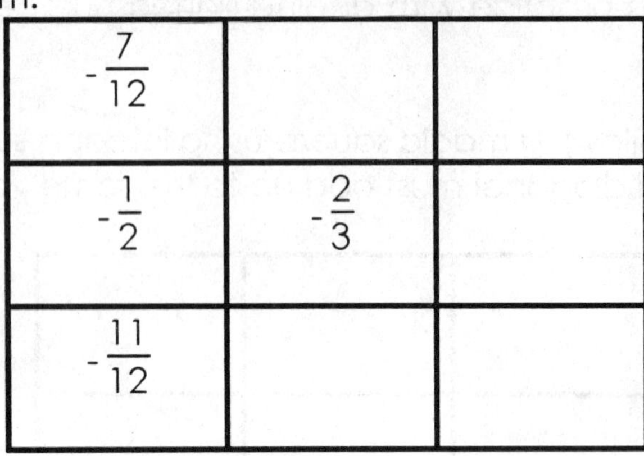

$-\frac{7}{12}$		
$-\frac{1}{2}$	$-\frac{2}{3}$	
$-\frac{11}{12}$		

The sum of each row, column, and diagonal is _______________.

4. a. Add the integers from -9 to -1 inclusive. The sum is ___________. Divide this sum by 9. The quotient is ___________. Does this number look familiar?

   b. Add the integers from -16 to -1 inclusive. The sum is ___________. Divide this sum by 16. The quotient is ___________. Does this number look familiar?

   c. The sum for a magic square can be found by ______________________________ and dividing by ___________________________________.

CONNECT THE COORDINATES

This activity provides practice plotting points in the Cartesian coordinate system.

Instructions: Draw the horizontal and vertical axes lightly on the graph paper. The horizontal axis should go from -8 to 8 and the vertical axis should range from -18 to 20. Plot the points and connect them in order, with line segments stopping when so indicated. If you have plotted all the points correctly, you should have a picture of a recognizable object when you are done.

1. (2, 0)
2. (3, 0)
3. (3.5, -1)
4. (3, -2)
5. (2, -3)
6. (2, -4)
7. (3, -5)
8. (4, -4)

STOP

9. (1, 15)
10. (1, -5)
11. (2, -7)
12. (2, -8)
13. (1, -10)
14. (0.5, -11)
15. (0.5, -15)
16. (0, -16)
17. (0, -18)
18. (0, -16)
19. (-3, -16)
20. (-5, -15)
21. (-7, -13)
22. (-8, -10)
23. (-8, -7)
24. (-6, -4)
25. (-5, -3)
26. (-6, -2.5)
27. (-5, -2)
28. (-4.5, -1)
29. (-5, 0)
30. (-6, 0.5)
31. (-5, 1)
32. (-6, 3)
33. (-6, 5)
34. (-5, 7)
35. (-3, 8.5)
36. (-1, 9)
37. (-1, 15)
38. (-1.5, 15)
39. (-1.5, 15.5)
40. (-1, 15.5)
41. (-1, 16)
42. (-1.5, 16)
43. (-1.5, 16.5)
44. (-1, 16.5)
45. (-1, 17)
46. (-1.5, 17)
47. (-1.5, 17.5)
48. (-1, 17.5)
49. (-1, 18)
50. (-1.5, 18)
51. (-1.5, 18.5)
52. (-1, 18.5)
53. (-1, 19)
54. (-2, 19)
55. (-3, 19.5)
56. (-2, 20)
57. (2, 20)
58. (3, 19.5)
59. (2, 19)
60. (1, 19)
61. (1, 18.5)
62. (1.5, 18.5)
63. (1.5, 18)
64. (1, 18)
65. (1, 17.5)
66. (1.5, 17.5)
67. (1.5, 17)
68. (1,17)
69. (1, 16.5)
70. (1.5, 16.5)
71. (1.5, 16)
72. (1, 16)
73. (1, 15.5)
74. (1.5, 15.5)
75. (1.5, 15)
76. (1, 15)

STOP

77. (-1, 9)
78. (-1, -5)
79. (-2, -7)
80. (-2, -8)
81. (-1, -10)
82. (- 0.5, -11)
83. (- 0.5, -15)
84. (0, -16)
85. (3, -16)
86. (5, -15)
87. (7, -13)
88. (8, -10)
89. (8, -7)
90. (6, -4)
91. (5, -3)
92. (6, -2.5)
93. (5, -2)
94. (4.5, -1)
95. (5, 0)
96. (6, 0.5)
97. (5, 1)
98. (6, 3)
99. (6, 5)
100. (5, 7)
101. (3, 8.5)
102. (1, 9)

STOP

103. (-2, 0)
104. (-3, 0)
105. (-3.5, 1)
106. (-3, -2)
107. (-2, -3)
108. (-2, -4)
109. (-3 -5)
110. (-4, -4)

STOP

NOTES

FISHING FOR LIKE TERMS

Today in class you will play a game called "Fishing For Like Terms" which will allow you to practice finding like terms. The rules of the game are as follows.

Game Rules:

1. One student shuffles the cards and deals five cards to each student.
2. The remaining cards are spread out face down in the center of the players. These cards are called the "FISH POND".
3. The person to the left of the dealer begins play by asking another player, by name, for cards which match a specific term held in his hand. For example, the player may say, "Jane, do you have any x squared y cubed cards?" If Jane has any x squared y cubed terms, no matter what the coefficient, she must give them all to the player who asked.
4. A player continues to ask other players for cards as long as he/she receives one or more cards from another player when asked.
5. If a player who is asked for a specific card has none, he tells the asking player to "FISH". The player asking for the cards then draws a card from the "FISH POND" and his turn is over. The player to his/her left then asks for a card.
6. When a player collects all four cards with matching terms, he/she shows the set to the other players and places the cards face down on the table.
7. If at any time during the game, a player runs out of cards, he/she immediately draws a card from the "FISH POND" and play continues.
8. The game is over when the last card is drawn from the "FISH POND".
9. The winner is the student who has the most sets of like terms at the end of the game.

WORKSPACE

FACTORING CROSSWORD

Complete the factoring puzzle on the next page by factoring each of the polynomials below. Each square should contain a number, a variable or an opening or closing parenthesis symbol. Remember that multiplication is commutative.

ACROSS

2. $x^2 + x - 20$
3. $3x^2 + 9x$
5. $2y^2 + 11y + 12$
7. $6m^2 - 19m + 15$
10. $n^2 - 3n - 2$
11. $25x^2 - 4$
13. $6r^2 + 19r - 7$
15. $1 - x^2$
16. $12y - 30$
17. $2t^2 - 5t - 12$
19. $8x^2 + 2x - 3$

DOWN

1. $2x^2 + 3x - 9$
2. $x^2 - 4$
4. $6p^2 + 35p + 25$
6. $6r^2 - 19r - 7$
8. $3x^2 + 5x - 2$
9. $3x^2 + 16x - 12$
12. $16r^2 - 1$
14. $9x^2 - 55x + 6$
18. $6r^2 + 23r + 20$

1
2
3
4
5
(y + 4)
6
7
8
(p + 5)
9
10
11
12
13
14
15
16
17
18
19

FACTORING BEE

You will compete as a member of a team in a factoring bee. Your instructor will tell you the type of factoring problems that will be used during the contest. The problems may factoring the difference of squares, sums or differences of cubes, and quadratic trinomials. Problems requiring you to remove a common factor or to use grouping may also be used.

To prepare for the competition, you may want to meet with your team and practice by trying as many factoring problems as possible.

The procedures and rules for the factoring bee are listed below.

PROCEDURES:

1. The students will be divided into teams of 3-5 students each
2. The instructor will provide a member of the first team with a factoring problem on a 3X5 card. The student will write the problem on the board and then will have a specified time limit to factor the problem.
3. If the student gets the problem correct, the first team receives 1 point and a member of the second team receives a new problem.
4. If the student from the first team can not factor it or factors it incorrectly, the same problem will be given to a member of the second who then will try to factor it. This will continue throughout the groups until the preset time limit for the game expires.
5. The team with the most points will be declared the winner.

RULES

1. Students at the board may not receive help from their team.
2. While a student is at the board, the other teams may work quietly as a group on the problem. If a member of their team is given the opportunity to factor the problem, the team may not help once the student is at the board.
3. No problems may be taken to the board written out on paper!

WORKSPACE

WHAT'S THE DIFFERENCE?

This activity provides an introduction to the effect of changing m and b on the graphs of linear equations of the form $y = mx + b$.

1. What do you think the graph of $y = x$ will look like? Write in complete sentences and give as much detail as possible.

2. Using your graphing calculator, graph the following three equations in the same standard viewing window.

   $y = x$ $\quad$ $y = x - 5$ $\quad$ $y = x + 4$

3. a. In what ways are the graphs of the three equations alike?

   b. In what ways are the graphs of the three equations different?

   c. What do you think the graph of $y = x - 2$ will look like?

   d. What do you think the graph of $y = x + b$ will look like?

4. Using your graphing calculator, graph the following three equations in the same standard viewing window.

$y = x$ $y = 2x$ $y = \frac{1}{3}x$

5. a. In what ways are the graphs alike?

b. In what ways are the graphs different?

c. What do you think the graph of $y = 5x$ will look like?

d. What do you think the graph of $y = mx$ will look like?

6. Using your graphing calculator, graph the following three equations in the same standard viewing window.

$y = -x$ $y = -2x$ $y = -\frac{1}{3}x$

7. a. In what ways are the graphs of the three equations alike?

   b. In what ways are the graphs of the three equations different?

   c. What do you think the graph of $y = -5x$ will look like?

   d. What do you think the graph of $y = mx$ will look like?

8. Using your graphing calculator, graph the following four equations in the same standard viewing window.

$y = 2x - 3$ $y = -2x - 3$ $y = -\frac{1}{3}x + 3$ $y = \frac{1}{3}x + 3$

9. a. In what ways are the graphs alike?

b. In what ways are the graphs different?

c. What do you think the graph of $y = -5x + 4$ will look like?

d. What do you think the graph of $y = mx + b$ will look like?

PEOPLE GRAPHING - LINES

This is a class activity that provides reinforcement on plotting points and graphing linear equations.

Sketch the lines created in class on the grids below. Answer the questions in complete sentences.

1. Equation: ____________
   Sketch:

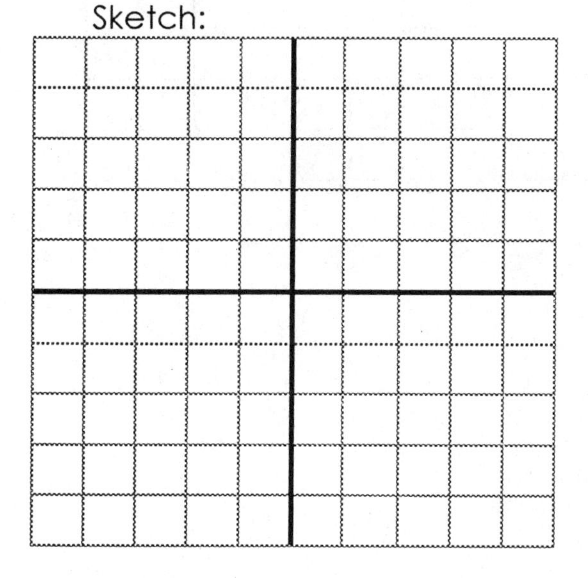

This is a ____________________ line.

2. Equation: ____________
   Sketch:

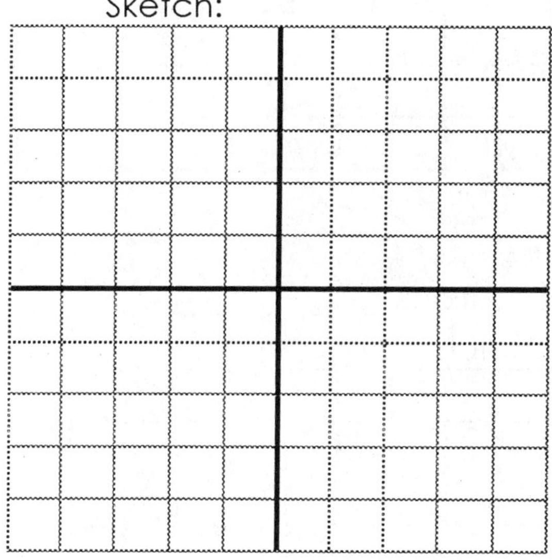

This is a ____________________ line.

3. Given the equation $y = -4$, describe the line. List at least two points on the line.

4. Given the equation $x = 6$, describe the line. Find two solutions of this equation.

5. Equation: 

Sketch:

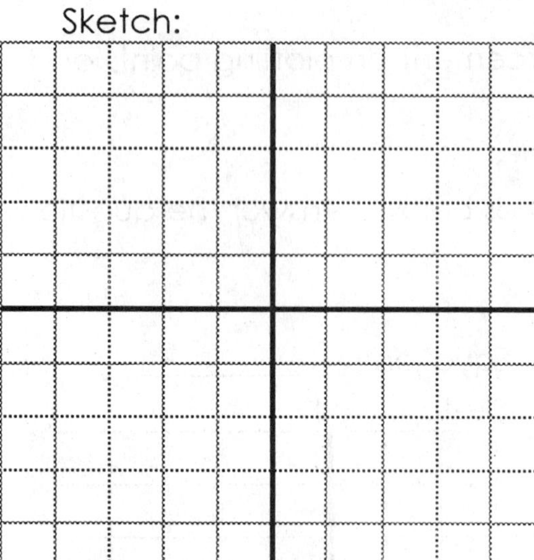

6. Equation: ____________

Sketch:

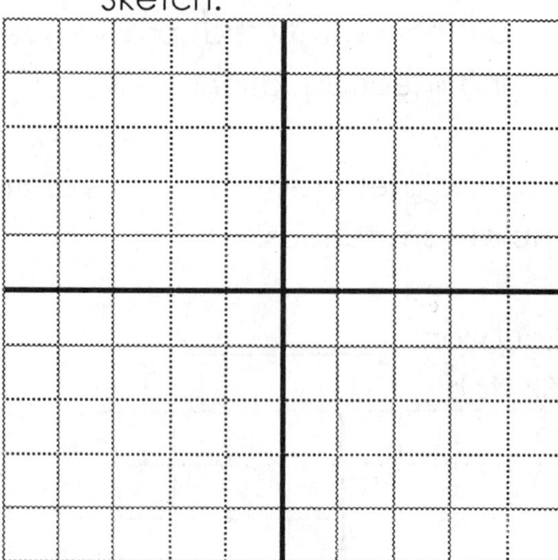

7. Equation: ____________

Sketch:

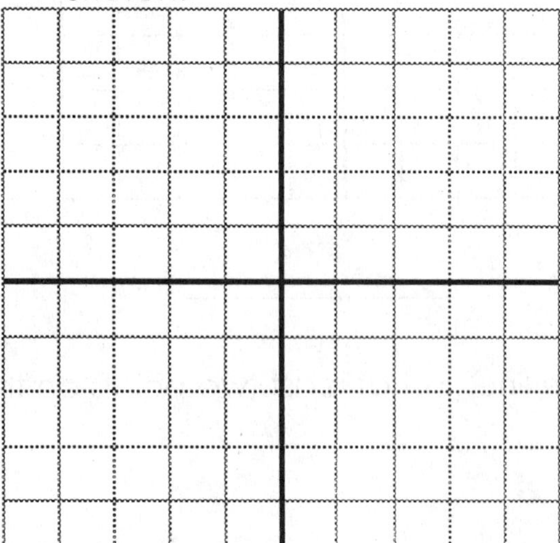

8. What effect does the coefficient of x have on the line? Describe, in complete sentences, how the line is affected.

9. Equation: ____________

Sketch:

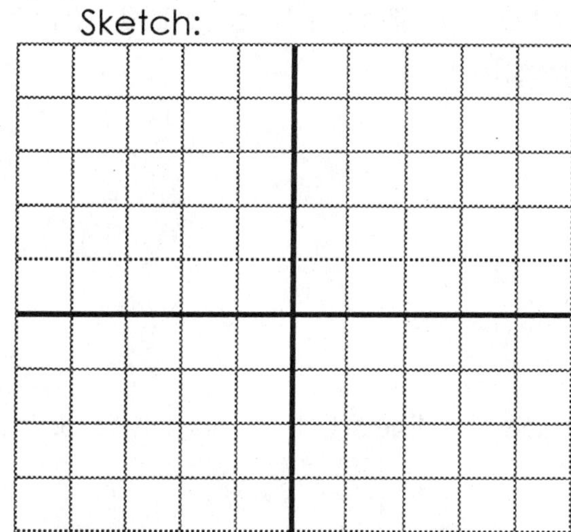

10. Equation: ______________

Sketch:

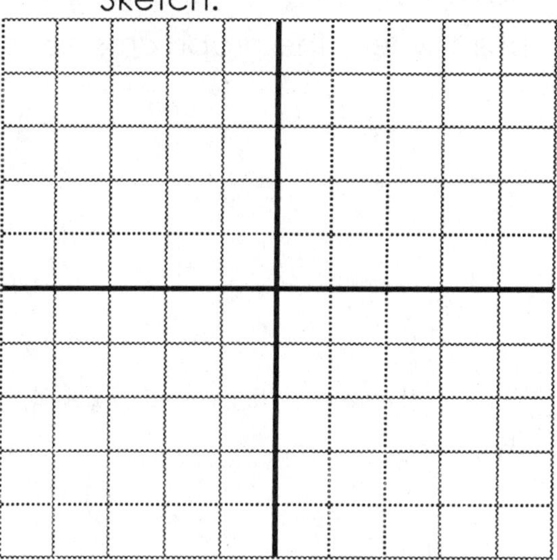

11. Compare, in complete sentences, the difference between lines with a positive coefficient of x and a negative coefficient of x. The coefficient is called the slope of the line.

12. Equation: ____________

Sketch:

13. Equation: ______________

Sketch:

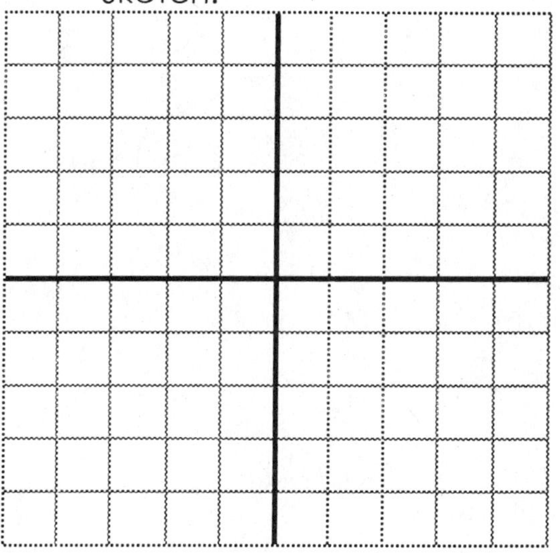

14. Describe, in complete sentences, what effect adding or subtracting a term has on the line. What is the y-intercept of the lines in 12 and 13? (This is the point where the graph crosses the y-axis.)

15. Given the equation $y = -2x + 4$, describe the line. List at least two points on the line.

16. Given the equation $y = 5x - 2$, describe the line. Find two solutions to the equation.

WHAT'S THE EQUATION OF MY LINE?

You will try to match the graph of an line to the equation of the line in this activity. Your instructor will tell you which version you will complete. To prepare for the activity, you should review how to identify the slope of a line from an equation and from a graph. You should also be able to find the vertical intercept from an equation and a graph. Practice changing any form of a linear equation into the slope-intercept form.

The rules for each of the versions of the activity are below. Be sure to bring this sheet to class.

VERSION 1

The object of this version is to match a graph of a line with the equation of a line.

RULES

a. All cards must remain turned over until the instructor says "start".
b. Students must match a graph to an equation, stacking the two cards.
c. When all matches are made the team leader raises his/her hand and says "stop'.
d. All groups must stop work immediately when the first team is finished.
e. The instructor will check the matches and if all are correct will award six points for that team. If not correct, all teams will continue work until all matches are made.
f. When one team has successfully completed all six matches, each of the other teams will be awarded one point for each of the matches that they have made.

VERSION 2

Using a graphing calculator, one student will enter a given equation and a specified window. The other student must find the equation of the line and demonstrate it on his/her calculator .

RULES

a. All cards must remain turned over until the instructor says "start".
b. The first student in each group will choose a card and enter the equation and the window given in the calculator.
c. The calculator is handed to the second student with the graph in the window.
d. The second student must find the equation of the line and enter it in his/her calculator and show that it matches the first graph to earn a point.
e. The second student may look at the window dimensions and may use the TRACE key.
f. Once the second student has decided that he/she has found the equation, the first student reveals the equation.
g. If the equation is correct, the second student receives a point. If not, no point is awarded.
h. The students now reverse roles, and repeat the procedure until time is called.

VERSION 3

Individual students will try to match the equation of a line with the graph of the equation.

HEIGHT VERSUS ARMSPAN

The relationship between a person's armspan and height can be expressed as a linear equation. Armspan is defined as the distance from the middle of a person's back to the tip of their fingers when their arm is held out to the side. Height is measured from head to foot.

1. Measure three of your classmate's armspan and height in inches and record the data in the chart below.

Name	Armspan in inches	Height in inches
1.		
2.		
3.		

2. Plot the three points on the graph below and connect two of the points in a line.

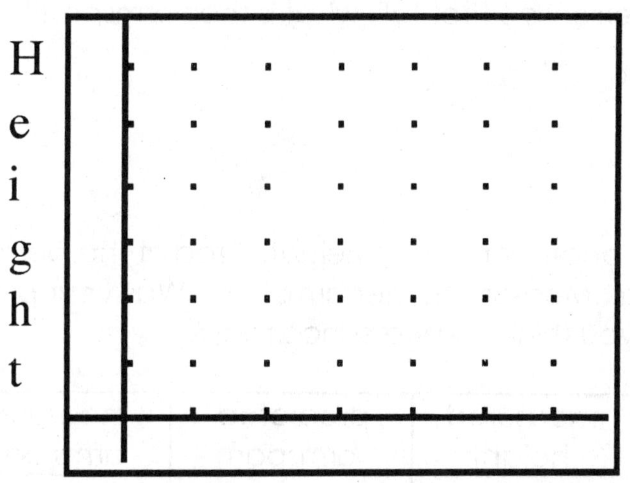

3. Using the two points you chose to connect above, calculate the slope of the line. What does the slope represent for this data?

4. Find an equation for the line. Put it in slope-intercept form.

5. Using your equation, find the following:
   A. How tall would a person be whose armspan is 30 inches?

   B. How tall would a person be whose armspan is 20 inches?

   C. If a person is 6 feet tall, what is their armspan?

6. Measure another person's height. Predict the person's armspan using your equation. Measure his/her armspan. Was your prediction correct? If not, why do you think you were incorrect?

measured height	predicted armspan	measured arms pan

WHAT'S MY INEQUALITY?

You will work in groups to write an inequality that will describe each of the graphs below. Complete as many inequalities on the worksheet below as your instructor requires and turn it in to your instructor.

INEQUALITY 1

a. Make a sketch of inequality 1 in the window below. Indicate the scale.

b. Find the slope and the vertical intercept of the line that borders the inequality and write the equation of that line.

c. Test several points that you can identify in the shaded area to see if the inequality should be written in the form: $y < mx + b$ or $y > mx + b$. Is the line bordering the shaded area a solid line or a dotted line? Use this information to complete your inequality.

d. The inequality describing the shaded area in Inequality 1 is:

____

INEQUALITY 2

a. Make a sketch of inequality 2 in the window below. Indicate the scale.

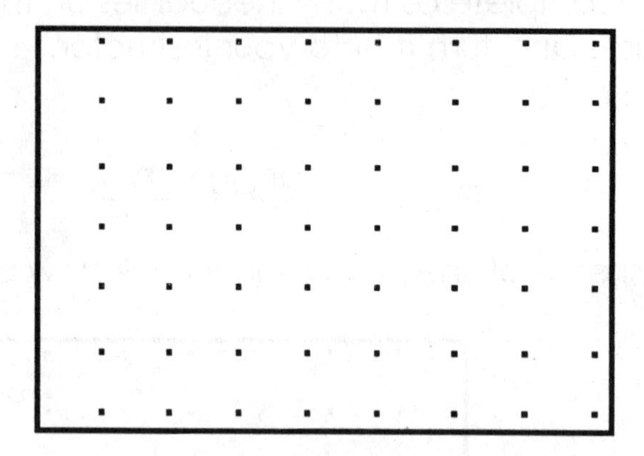

b. Find the slope and the vertical intercept of the line that borders the inequality and write the equation of that line.

c. Test several points that you can identify in the shaded area to see if the inequality should be written in the form: $y < mx + b$ or $y > mx + b$. Is the line bordering the shaded area a solid line or a dotted line? Use this information to complete your inequality.

d. The inequality describing the shaded area in Inequality 2 is:

____

INEQUALITY 3

a. Make a sketch of inequality 3 in the window below. Indicate the scale.

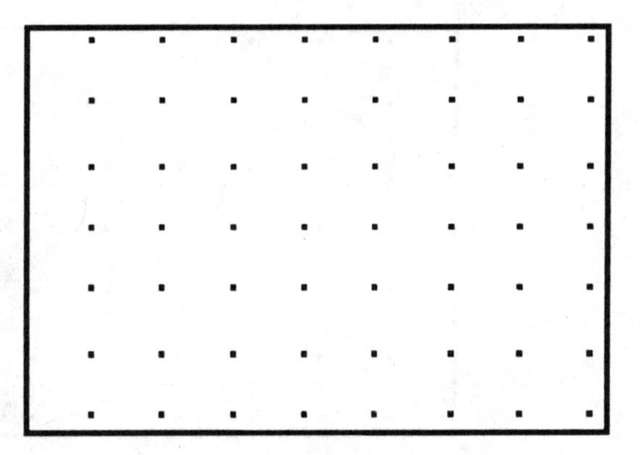

b. Find the slope and the vertical intercept of the line that borders the inequality and write the equation of that line.

c. Test several points that you can identify in the shaded area to see if the inequality should be written in the form: $y < mx + b$ or $y > mx + b$. Is the line bordering the shaded area a solid line or a dotted line? Use this information to complete your inequality.

d. The inequality describing the shaded area in Inequality 3 is:

____

INEQUALITY 4

a. Make a sketch of inequality 4 in the window below. Indicate the scale.

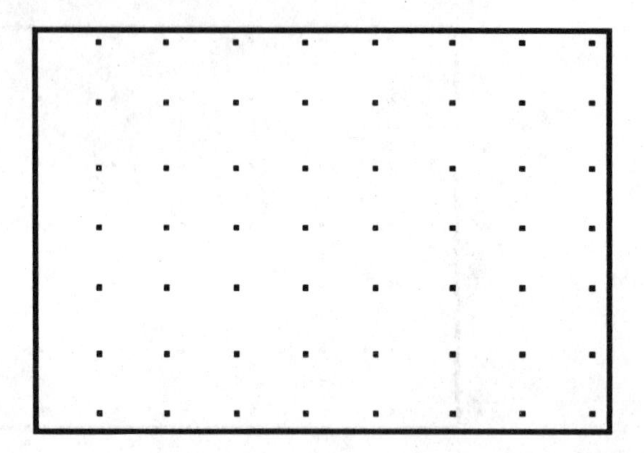

b. Find the slope and the vertical intercept of the line that borders the inequality and write the equation of that line.

c. Test several points that you can identify in the shaded area to see if the inequality should be written in the form: $y < mx + b$ or $y > mx + b$. Is the line bordering the shaded area a solid line or a dotted line? Use this information to complete your inequality.

d. The inequality describing the shaded area in Inequality 4 is:

____

INEQUALITY 5

a. Make a sketch of inequality 5 in the window below. Indicate the scale.

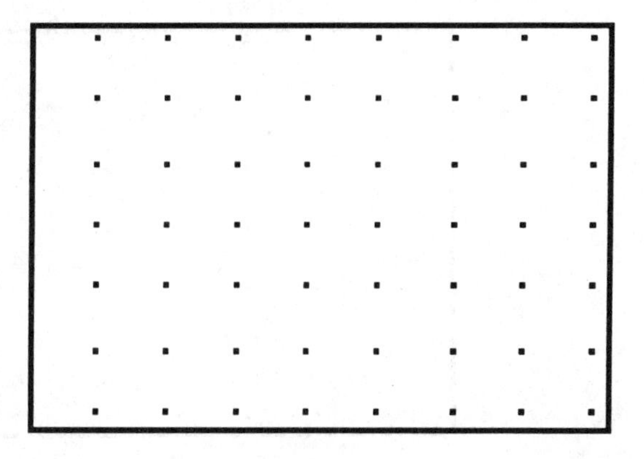

b. Find the slope and the vertical intercept of the line that borders the inequality and write the equation of that line.

c. Test several points that you can identify in the shaded area to see if the inequality should be written in the form: $y < mx + b$ or $y > mx + b$. Is the line bordering the shaded area a solid line or a dotted line? Use this information to complete your inequality.

d. The inequality describing the shaded area in Inequality 5 is:

____

INEQUALITY 6

a. Make a sketch of inequality 6 in the window below. Indicate the scale.

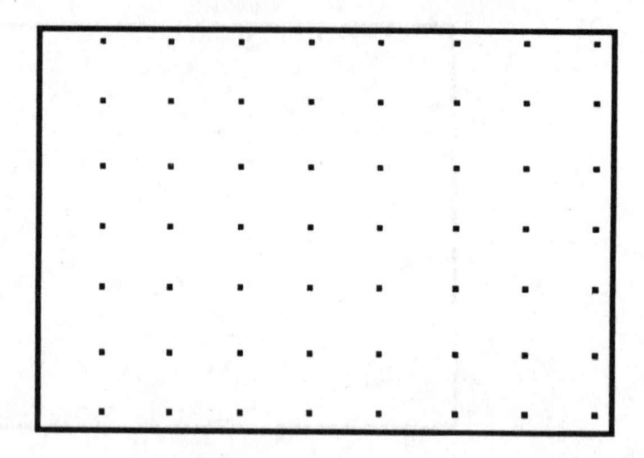

b. Find the slope and the vertical intercept of the line that borders the inequality and write the equation of that line.

c. Test several points that you can identify in the shaded area to see if the inequality should be written in the form: $y < mx + b$ or $y > mx + b$. Is the line bordering the shaded area a solid line or a dotted line? Use this information to complete your inequality.

d. The inequality describing the shaded area in Inequality 6 is:

____

PERPENDICULAR LINES

There is a particular relationship between the equations of two perpendicular lines. In this activity, you will discover that relationship and use it to predict the equation of one of the perpendicular lines if you know the equation of the other perpendicular line.

You will work in groups of 2-3 students to complete the activity sheet and then the class will discuss the results.

Part 1

a. Plot the points (1,3) and (4,9) on a sheet of graph paper. Be sure to label the axes and indicate the scale.
b. Draw a line that includes both the points using a ruler.
c. Determine the slope of the line. (You may find the slope using the graph or using an equation.)

d. Find the y-intercept of the line using the graph or an equation.

e. Write the equation of this line in the form $y = mx + b$. (Write this equation beside the line on your graph paper also.)

f. Using the method demonstrated by your instructor, draw a line perpendicular to your first line through the point (3,7). Be sure to extend the line so that it intersects the vertical-axis.
g. Find the slope of the perpendicular line.

h. Find the y-intercept of the line.

i. Write the equation of the line and label its graph.
j. Repeat steps f-i for the point (2,5).

k. What relationship, if any, do you see between the slope of your original line and the slope of each of the perpendicular lines?

l. What relationship, if any, do you see between the y-intercepts of the original line and the slope of each of the perpendicular lines?

PART 2

a. Plot the points (– 3, 1) and (3, – 3) on a new sheet of graph paper. Be sure to label the axes and indicate the scale.
b. Draw a line that includes both the points using a ruler.
c. Determine the slope of the line. (You may find the slope using the graph or using an equation.)

d. Find the y-intercept of the line using the graph or an equation.

e. Write the equation of this line in the form $y = mx + b$. (Write this equation beside the line on your graph paper also.)

f. Using the method demonstrated by your instructor, draw a line perpendicular to your first line through the point (0, – 2).
g. Find the slope of the perpendicular line.

h. Find the y-intercept of the line.

i. Write the equation of the line and label its graph.

j. What relationship, if any, do you see between the slope of your original line and the slope of the perpendicular line?

k. What relationship, if any, do you see between the y-intercepts of the original line and the slope of the perpendicular line?

l. Compare the relationships found in Part 1 and Part 2.

PART 3

a. Draw the graph of the line $x = 3$ on a new sheet of graph paper.
b. What is the slope of this line?
c. Does this line have a y-intercept? If so, what is it?

d. Does this line have an x-intercept? If so, what is it?

e. Draw a line perpendicular to x = 3 through the point (3,7).

f. What is the slope of this line?

g. Does this line have a y-intercept? If so, what is it?

h. Does this line have an x-intercept? If so, what is it?

i. What is the equation of this line?

j. What relationship do you see between the slopes of the two lines? Is it the same as the relationship between the slopes in Part 1 and Part 2. If not, why not?

PART 4

a. If you know that the equation of a line is $y = \frac{2}{3}x - 1$, find the equation of a perpendicular that will pass through the point (3,1).

b. A horizontal line has the equation y = – 5. Write the equation of a perpendicular line that passes through the point (2, – 5).

c. The slope-intercept form of the equation of a line is y = mx + b. How could you represent the slope of a line perpendicular to this line?

PEOPLE GRAPHING - INEQUALITIES

This is a class activity that provides reinforcement on graphing linear inequalities. This activity provides visualization of solution sets for linear inequalities.

Sketch the lines and shaded regions created in class on the grids below.

1. Inequality: ____________
   Sketch:

2. Inequality: ____________
   Sketch:

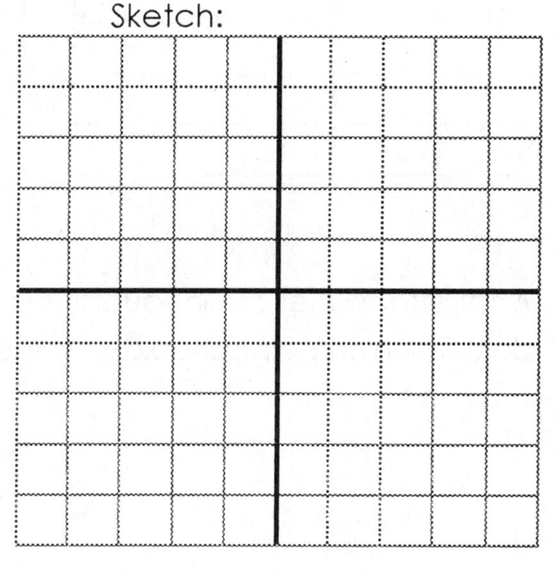

3. Given the inequality $y \leq -4$, describe the graph of the solution set.

4. Given the inequality $x > 3$, determine whether the points below are in the solution set. How did you determine which were solutions?

Point	Is it a solution?
(-1, 7)	
(5, 2)	
(3, -5)	
(4, 0)	
(0, -8)	

5. Inequality: ____________

Sketch:

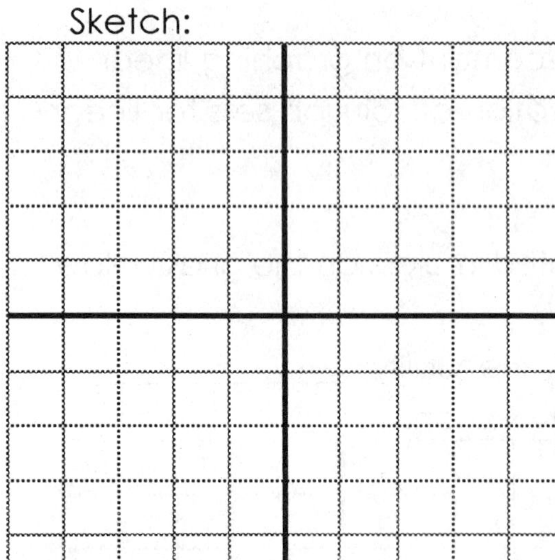

6. Inequality: ______________

Sketch:

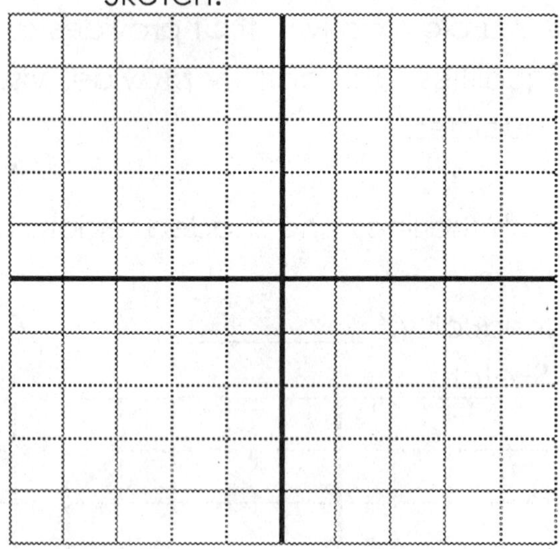

7. What effect does the inequality symbol have on the graph? How did you decide which points should be shaded?

8. Inequality: ____________

Sketch:

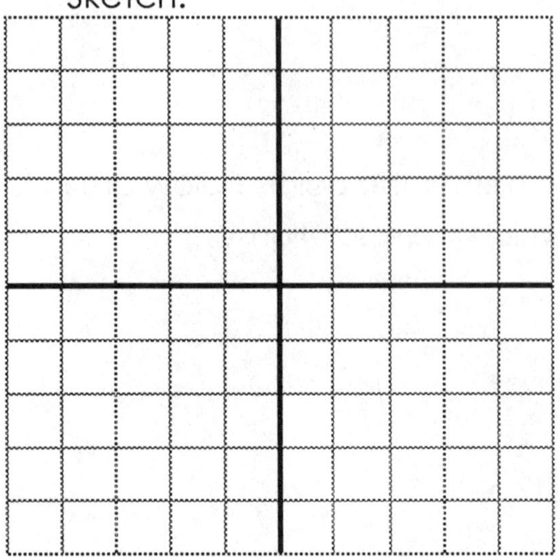

9. Inequality: ______________

Sketch:

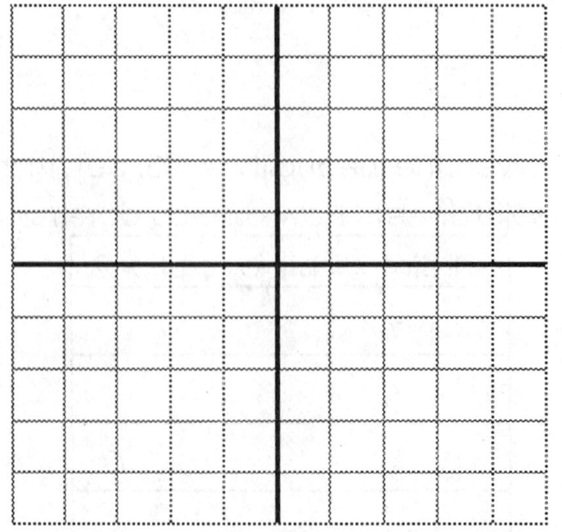

10. Compare, in complete sentences, the difference between the graphs in 8 and 9. Explain how you determined what properties the graph would have. How did you know whether the line was solid or dotted? How did you determine which side would be shaded?

11. Inequality: ____________

Sketch:

12. Inequality: ____________

Sketch:

13. List the steps you would take to make a complete graph of the solutions to the inequality $y < 2x + 1$.

WORKSPACE

PEN AND INK

Suppose that your supervisor at work puts you in charge of getting the weekly flyer printed. There are two print shops in town, PEN and INK, and you call each shop for a quote. At PEN the quote is a $50 set-up fee plus $0.025 per page printed and at INK the quote is a $30 set-up fee plus $0.04 per page printed. From week to week the number of pages that you will need printed varies.

1. Complete the following table for the costs of printing based on the quotes above.

Number of pages	Cost at PEN	Cost at INK
500		
1000		
1500		
2000		
x		

2. Graph the equation for the cost at PEN and the equation for the cost at INK on the grid below.

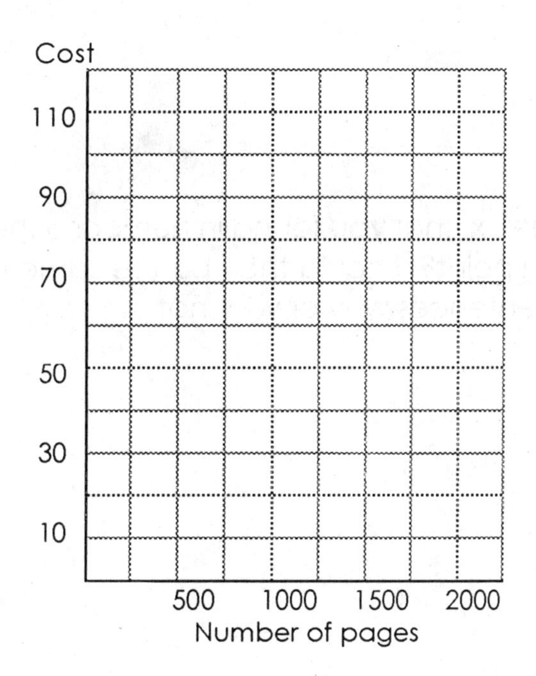

3. Find the intersection point of the two graphs. What is the x-value of this point? What is the y-value of this point? Explain in complete sentences what this point represents in terms of the number of pages and the cost at the two shops.

x = ______ y = ____

4. Write an equation for the cost at PEN. Cost at PEN = ______________.

Write an equation for the cost at INK. Cost at INK = ______________.

5. To find the number of pages for which both PEN and INK will charge you the same price, you need to set the two equations in number 4 equal to each other. Set the equations equal and solve for x.

6. Is the number of pages, x, that you found in number 5 the same as the x-value of the intersection point? Should they be the same number? Explain in complete sentences why or why not.

PEOPLE GRAPHING - LINEAR SYSTEMS

This is a class activity that provides reinforcement on graphical solutions to 2 by 2 systems of linear equations.

Sketch the lines created in class on the grids below. Answer the questions in complete sentences.

1. Equation 1: ____________

   Equation 2: ____________

   Sketch:

   The solution is ____________________.

2. Equation 1: ____________

   Equation 2: ____________

   Sketch:

   The solution is ____________________.

3. Substitute the solution to system #1 into both equations and verify that the solution makes both equations true.

4. Equation 1: ____________

   Equation 2: ____________

   Sketch:

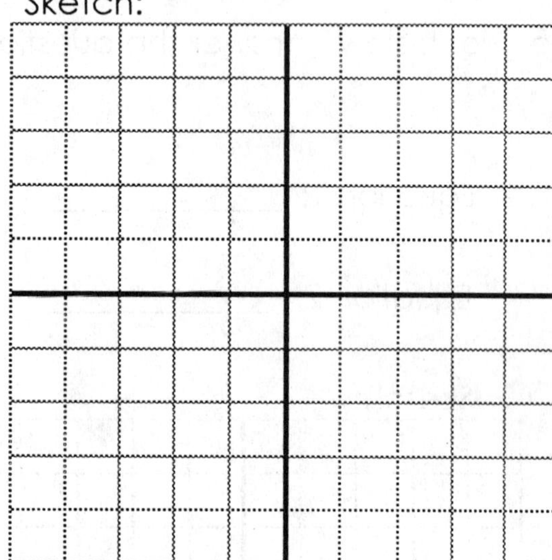

   The solution is ____________________.

5. Equation 1: ____________

   Equation 2: ____________

   Sketch:

   The solution is ____________________.

6. Explain in complete sentences, how to find the solution to a system of linear equations.

7. Equation 1: ____________

Equation 2: ____________

Sketch:

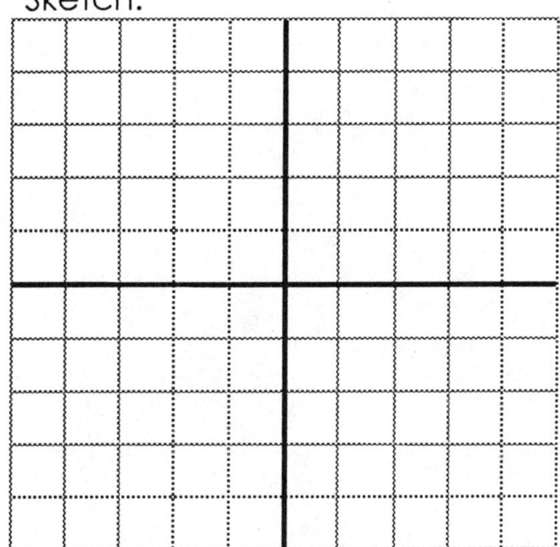

The solution is ____________________.

8. Equation 1: ____________

Equation 2: ____________

Sketch:

The solution is ____________________.

9. An inconsistent system is a system which has no solutions. What kind of lines are in an inconsistent system?.

10. Equation 1: ____________

Equation 2: ____________

Sketch:

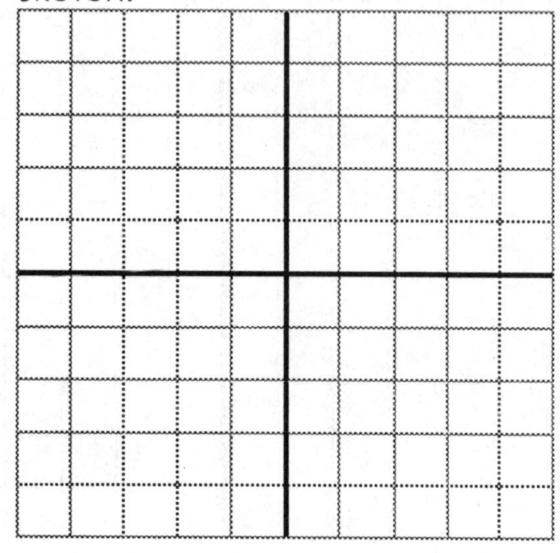

11. The system in #10 is a dependent system. What is the solution set for this problem? Describe the relationship between the lines.

RING AROUND THE ROOM

This activity requires you to measure the classroom and calculate the perimeter. You will also determine how many packages of wallpaper border would be required to decorate the classroom. Be sure to show all calculations for the following questions.

1. Measure the dimensions of this room and sketch a picture of this room on the grid below.

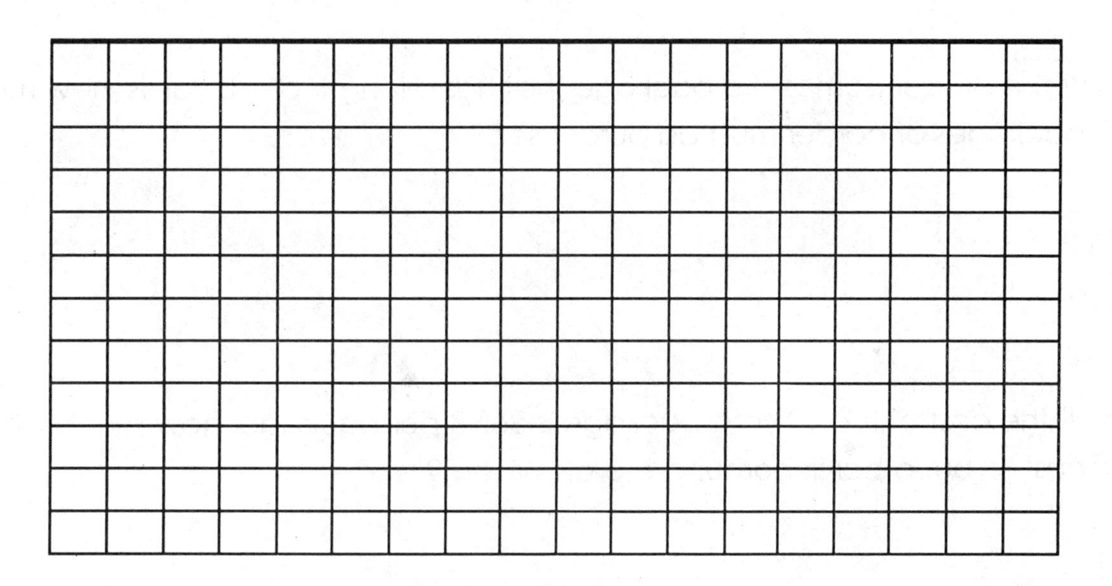

2. The perimeter of this room measures ________ feet.

3. Instead of feet, what other measuring units could you have used?

4. The perimeter of this room measures ________ yards.

You wish to hang a wallpaper border on each of the walls. The border you have chosen comes in packages that measure 4 inches high by 16 feet long.

5. How many packages of the border must be purchased to go around the room?

6. If the border comes in a package that has a length of 10 yards, how many packages of border must be purchased?

7. If the cost of the 10 yard package is $6.98 per package, how much will it cost to put a border around the classroom?

ATTRIBUTE MATCH

This activity provides practice finding the concavity, vertical intercept, line of symmetry and vertex of parabolas from their equations. For each of the following quadratics, tell whether the parabola opens upward or downward, give the vertical intercept, give the equation for its line of symmetry, identify the vertex, and sketch a graph of the parabola using these attributes. At your next class meeting, you will be asked to match graphs of parabolas with their equations so be sure you understand how to do these problems.

1. $y = -2x^2 + 8x - 4$

 a. This parabola opens ______________________ because

 b. The vertical intercept of this parabola is (,)

 c. The vertex of this parabola is (,)

 d. The equation for the line of symmetry of this parabola is:

 e. The graph of this parabola looks like:

2. $y = 2(x - 3)^2 + 6$

a. This parabola opens ____________________ because

b. The vertex of this parabola is (,   )

c. The equation for the line of symmetry of this parabola is:

d. The vertical intercept of this parabola is (,   )

e. The graph of this parabola looks like:

3. $y = x^2 - x - 2$

a. This parabola opens ____________________ because

b. The vertical intercept of this parabola is (,   )

c. The vertex of this parabola is (,   )

d. The equation for the line of symmetry of this parabola is:

e. The graph of this parabola looks like:

4. $y = 6(x - 2)^2 + 3$

a. This parabola opens ____________________ because

b. The vertex of this parabola is (,)

c. The equation for the line of symmetry of this parabola is:

d. The vertical intercept of this parabola is (,)

e. The graph of this parabola looks like:

5. $y = 3x^2 - 5x + 1$
   a. This parabola opens ____________________.

   b. The vertical intercept of this parabola is (,)

   c. The vertex of this parabola is (,)

   d. The equation for the line of symmetry of this parabola is:

   e. The graph of this parabola looks like:

6. $y = 3(x + 4)^2 - 12$
   a. This parabola opens ____________________.

   b. The vertex of this parabola is (,)

   c. The equation for the line of symmetry of this parabola is:

   d. The vertical intercept of this parabola is (,)

   e. The graph of this parabola looks like:

7. $y = 2x^2 - x - 10$

   a. This parabola opens ______________________ .

   b. The vertical intercept of this parabola is (,)

   c. The vertex of this parabola is (,)

   d. The equation for the line of symmetry of this parabola is:

   e. The graph of this parabola looks like:

8. $y = -(x + 2)^2 - 4$

   a. This parabola opens ______________________ .

   b. The vertex of this parabola is (,)

   c. The equation for the line of symmetry of this parabola is:

   d. The vertical intercept of this parabola is (,)

   e. The graph of this parabola looks like:

9. $h = -16t^2 + 64t + 4$

   a. This parabola opens ____________________.

   b. The vertical intercept of this parabola is (,)

   c. The vertex of this parabola is (,)

   d. The equation for the line of symmetry of this parabola is:

   e. The graph of this parabola looks like:

10. $h = -16(t-2)^2 + 68$

   a. This parabola opens ____________________.

   b. The vertex of this parabola is (,)

   c. The equation for the line of symmetry of this parabola is:

   d. The vertical intercept of this parabola is (,)

   e. The graph of this parabola looks like:

OFF TO WORK WE GO

In this activity, you will be participating in a "real-life" work problem. This activity should not be a race! Try to maintain a constant speed when placing the beads in the container. One person in your group should be the timekeeper. The other two people will be doing the work.

1. Have the first worker place all the beads, one at a time, into the container using only one hand.

   It took the first worker _______ seconds to put all the beads in the container.

2. Have the second worker place all the beads, two at a time, into the container using only one hand.

   It took the second worker _______ seconds to put all the beads in the container.

3. Have the workers work together to place all the beads into the container. Each worker should use the same method and speed as before.

   It took both workers _______ seconds to put all the beads in the container.

4. Who took the longest to complete the task? Circle the correct answer.

   worker 1 worker 2 both workers together

5. Who completed the task in the shortest time? Circle your answer.

   worker 1 worker 2 both workers together

6. The work equation says:

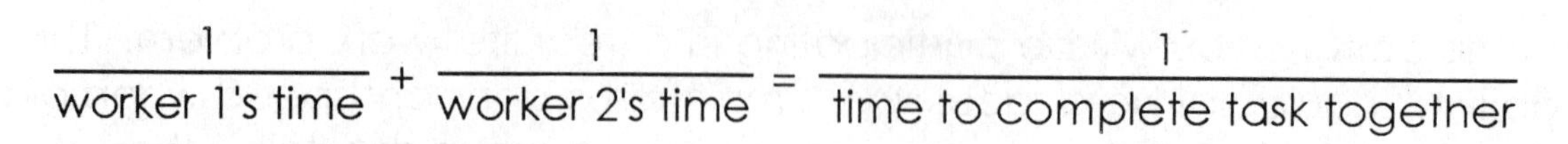

$$\frac{1}{\text{worker 1's time}} + \frac{1}{\text{worker 2's time}} = \frac{1}{\text{time to complete task together}}$$

Substitute the times you found for worker one and worker two into the above equation and solve for the time it should take you to complete the task when you worked together.

7. Compare the time it took you to put the beads in the container when you worked together with the time the equation predicts. Were they the same? Why or why not?

GOOD TASTING RATIOS

This activity provides practice with ratios and proportions.

Open your box of M&M's® and pour them onto a sheet of paper. (NO nibbling!)

1. Sort by color and count the number of each color. Fill in the table below.

Color	Number of M&M's® in your box	Number of M&M's® in another box	Number in both boxes	Ratio (number/total) for both boxes
Green				
Orange				
Pink				
Yellow				
Blue				
Brown				
Red				
Total for the bag				

Looking at the number of each color in both boxes and answer the following questions.

2. a. Which color has the largest number of M&M's®? What is the ratio?

   b. Which has the least? What is the ratio?

When comparing your box to the other box:

3. a. Did each box have the same number of M&M's®? How many did each have?

   b. What is the average number of M&M's® per box?

4. Find the following ratios:
   a. red to blue

   b. pink to blue

   c. red to orange

   d. Do you believe that the M&M® company fills the boxes from a mix that has the same ratio of each color? Explain your answer in complete sentences.

5. The M&M® Company reports that (as of 10-96) the production ratio for plain mini M&M's® is as follows:

Color	Percent
Green	18.75
Orange	12.5
Yellow	12.5
Blue	18.75
Brown	12.5
Pink	12.5
Red	12.5

a. Do your boxes fit this ratio?

b. Given the above ratios, the number of blue M&M's® should be 18.75 per 100, set up and solve a proportion to estimate how many M&M's® will be blue in a total of 575 M&M's®.

c. Given the above ratios, set up and solve a proportion to estimate how many M&M's® will be pink in a total of 250 M&M's®.

WORKSPACE

TO AND FRO

The purpose of this activity is to explore the relationship between the length of a pendulum and the time it takes to swing back and forth (this time is called the period). You will vary the length of a pendulum, L, and record the corresponding period, T. To record the period, find the time it takes for the pendulum to make 10 full swings back and forth and then divide that time by 10 to find the time for one period. You will then fit a curve to this data. This curve will be used to predict how long a pendulum should be in order to make the period be 1 second.

1. Record your data in the table.

Length, L (cm)	Period, T (sec)	Length, L (cm)	Period, T (sec)
0	0	100	
10		110	
20		120	
30		130	
40		140	
50		150	
60		160	
70		170	
80		180	
90		190	

2. Graph your points on the grid below.

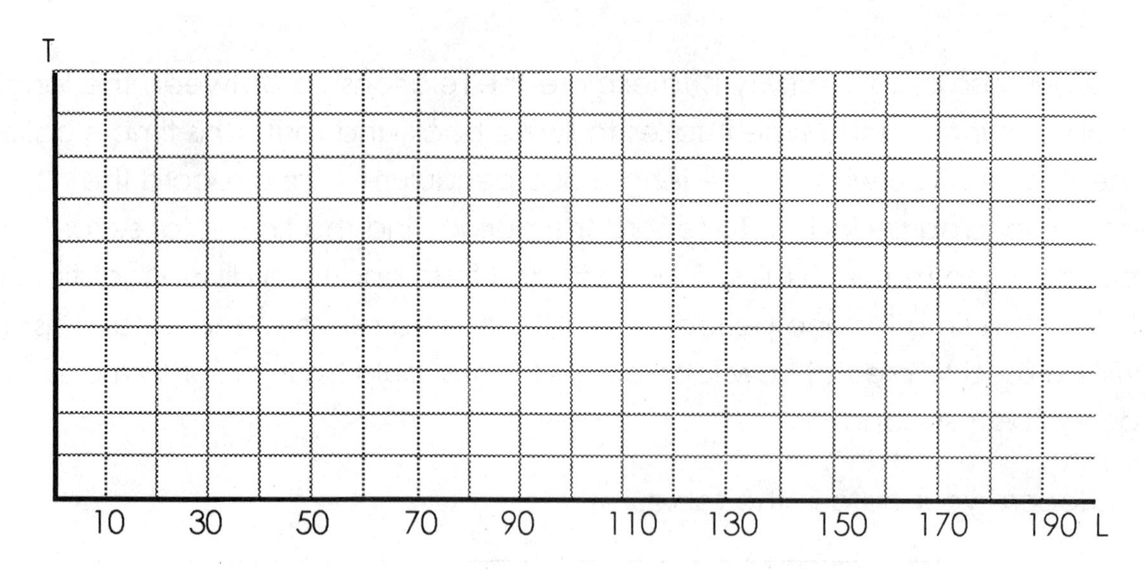

This graph should look like a square root graph. To find the coefficient and exponent for the data, you will need a system of equations. The general form of the equation is $T = aL^b$. You need to find the coefficient, a, and the exponent, b.

3. Using the points corresponding to L = 10 and L = 100, a system of equations that could be solved for the constants a and b is:

4. If you square both sides of the equation in which L = 10, you can solve for the constants. Show all work to find the constants.

5. When a and b are substituted into the general form, an equation for the data collected is ________________. (The exponent should be approximately 0.5)

6. Graph the data and your equation for the data below.

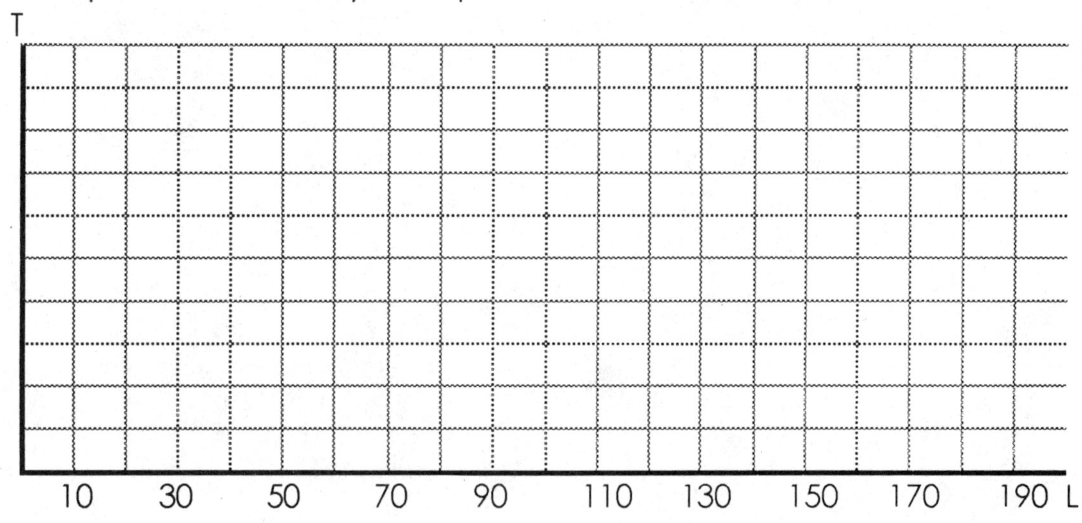

7. How well does the equation appear to fit the data?

8. Using your equation, calculate the length of the pendulum that would have a period of one second. Show your calculations below.

9. Test your prediction above. What was the period of the pendulum with the length you found in 8?

WORKSPACE

A PAINTING FUNCTION

If you can develop a function to calculate the cost of paint that would be required to paint a wall, then the function could be modified and used for finding the cost of a room or an entire building. Work with a partner to complete this worksheet.

1. Find the area of one of the walls in your classroom in square feet. Choose a wall that has no doors, windows or blackboards if possible. If you do not have a wall with openings or blackboards, ignore the disruptions.

   HEIGHT OF WALL: ____________________

   LENGTH OF WALL: ____________________

   AREA OF WALL: ____________________

   CALCULATIONS:

2. In a recent newspaper ad, a gallon of top quality interior paint sold for $21.88. A gallon of paint could cover 200 square feet with a single coat of paint.

   a. Calculate the cost of the paint per square foot. Report the cost to 4 decimal places.

      COST OF PAINT: ____________________

      CALCULATIONS:

   b. Calculate the cost of painting your classroom wall.

      COST OF PAINTING THE WALL: ____________________

      CALCULATIONS:

3. If you needed to calculate the cost for many walls, you could repeat steps 1 and 2 above. Or, you could develop a function that could be used for any wall. Complete the following steps to find a function of a wall with no interruptions.

   a. Assume that the height of all the walls you will be working with is the same as the height of your classroom, so you will use that height in all calculations.

      HEIGHT OF ALL WALLS: ______________________________

   b. Assume that the length of each wall may vary, and choose a variable to represent that length.

      LENGTH OF ANY WALL: ______________________________

   c. Write an expression for the area of any wall using the information in a and b above.

      AREA OF ANY WALL: ______________________________

   d. Assume that you will purchase the paint used in part 2. Write an expression that you can use to calculate the total cost of putting one coat of paint on the wall described in c.

      COST OF PAINTING ANY WALL: ______________________________

4. a. To write a function, you should choose a letter that will represent the rule used for calculating the cost of painting the wall. You may use a capital letter or a lower case letter. You may use any letter of the alphabet, but you may want to choose one that would remind you what the function represented. For example, a function describing the time required for an object to fall from a roof might be described using the letter, t.

      FUNCTION RULE: ______________________________

   b. Now write the function representing the cost of painting a wall using the function rule from 4a, the variable from 3b, and the expression from 3d.

      FUNCTION DESCRIBING THE COST FOR PAINTING ANY WALL:

____

5. Apply the function you wrote in part 4 to calculate cost.

   a. You need to paint a second wall in your classroom that is 27 feet long. How much will it cost to put on one coat of paint?

   b. You have a helper who measured the perimeter of a room, but did not measure each individual wall. The perimeter is 122 feet. Assuming that the height of this room is the same as your original classroom, what would be the cost of putting one coat of paint on the walls?

6. Apply the function you wrote in part 4 to calculate length.

   a. You have bought 500 gallons of the paint described in part 2. Use your function to calculate the total length of wall that can be painted assuming that the walls are of the same height as your classroom.

   b. You have $218.80 worth of paint available. You have been asked to paint a large room with the width of 40 feet and the length of 80 feet. The height of the room is the same as your classroom. Do you have enough paint to put one coat on all the walls in the room? Justify your answer with calculations.

7. Rewrite your function from 4b for paint that is on sale for $15.98 per gallon assuming that this paint will also cover 200 square feet per gallon.

8. Rewrite your function from 4b if the height of all walls are 15 feet.

9. In a real life situation, the number of doors, windows, etc. will affect the amount of paint necessary to complete the job.

   a. Would you need more or less paint to complete the job if there were one or more openings in the walls of a room? Why?

   b. If there was a door with dimensions 3 feet by 8 feet in a classroom wall that is 26 feet long, by how much would the cost change? Show your calculations.

WHAT'S MY EXPRESSION?

You will try to write an algebraic expression if you are given an English phrase or the English phrase if you are given an algebraic expression. Your instructor will tell you which version of the exercise you will do. To prepare for the activity, you should review reading, writing and stating equivalent algebraic expressions and English phrases. For example, the English expression, the sum of three times a number and 6, will translate to $3x + 6$, $3n + 6$, or $3y + 6$, etc.

The rules for each of the versions of the activity are below. Be sure to bring this sheet to class.

VERSION 1

The object of this version is to write the algebraic expression from the English phrase read by a team member.

RULES

a. One student deals the "Expression" cards face down in front of the other members of the team until each member has at least 5 cards. (More may be dealt, but be sure each member of the team has the same number of cards.) Put the extra cards away.
b. The dealer chooses the top card on his stack, and reads the English phrase.
c. The team member to the dealer's right, attempts to write down the algebraic expression equivalent to the English phrase. The variables may differ from those on the cards. The team member may ask for the phrase to be repeated up to three times.
d. The team member shows the algebraic expression to the dealer. If it is correct, the team member earns a point.
e. If the team member does not write the algebraic expression correctly, the student to his/her right has the opportunity to write the algebraic expression. Again, if this team member answers correctly, he/she will receive a point.

f. If no team member is able to write the expression correctly, the dealer shows the card to all members and then the card is removed from play.
g. The team member to the right of the dealer now chooses his top card, reads the English phrase to the team member on the right and play continues as before.
h. Play continues until all cards have been read.

VERSION 2

One student from a team will write either an algebraic expression or an English phrase on the board or overhead projector and all members of the other teams will write either the English phrase of the algebraic expression as required.

RULES

a. All cards must remain turned over until the team is called on by the instructor.
b. A student from the identified team chooses the top card and writes either the English phrase or the algebraic expression on the board (or on a transparency on an overhead projector).
c. All students on the other teams write a corresponding algebraic expression or English phrase. (The variables may differ from those on the cards.)
d. The members of the team that presented the problem will check the other team members work.
e. The team presenting the problem will receive 1 point for every correct answer while the other teams will lose 1 point for every incorrect answer from members of their teams.
f. Play moves to the next team and points are accumulated as before.
g. Play will continue until each team has been allowed to present an equal number of times.

LET'S DECORATE THE CLASSROOM

This activity requires you to measure the classroom and calculate the area. You will determine how many gallons of paint would be needed to paint the classroom and how much carpet would be needed for the floor. Be sure to show all calculations needed to answer the following questions.

1. Measure the dimensions of the classroom in feet and record the values below.

   Length = ______ feet Width = ______ feet Height = ______ feet

2. The area of the right wall measures ________ square feet. The total area of the right and left walls is ________ square feet.

3. The area of the front wall measures _________ square feet. The total area of the front and rear walls is __________ square feet.

4. The total area for the walls is ___________ square feet.

5. The label on a gallon of latex paint estimates that the paint will cover about 400 square feet. How many gallons must be purchased for one coat of paint?

6. The area of the floor measures ________ square feet.

7. If carpeting costs $8.95 per square foot, how much will it cost to carpet the floor?

HOW LONG WILL YOU LIVE?

This activity requires you to measure the classroom and calculate the volume. You will determine how long the air would last if the room was sealed. Be sure to show all calculations needed to answer the following questions.

1. Measure the dimensions of the classroom in feet and record the values below.

   Length = ______ feet   Width = ______ feet   Height = _____ feet

2. The volume of the room measures _______ cubic feet.

3. One cubic foot is approximately 28, 505.11 cubic centimeters. Find the volume in cubic centimeters.

If the room were suddenly sealed and the entire class was in the room and there was no additional air, you will now calculate how long the air will last.

4. There are _______ people in the room. The average person breathes 12 breaths a minute. At this rate, how many total breaths per minute are being taken?

5. One milliliter equals one cubic centimeter. What is the volume of air in the room in milliliters?

6. With each breath, a person uses 350 milliliters of air. How much total air is used by one person in one minute?

7. How much total air is used by the class in one minute?

8. The air will last for _________ minutes. The air will last for ________ hours.

9. When people are nervous, they breathe faster. Predict the effect if everyone breathes 14 breaths per minute.

WHAT'S IN THE BAG?

This activity provides an introduction to bar graphs, practice with estimation and the calculation of percents.

1. Open your bag of M&M's® and pour them onto a sheet of paper. (NO nibbling!)
2. Sort by color and count the number of each color. Fill in the table below.

Color	Number of M&M's®	Percent (number/total)
Green		
Orange		
Yellow		
Blue		
Brown		
Red		
Total for the bag		100%

3. Graph each color total on the graph below. The colors are on the horizontal axis and the number of M&M's® are on the vertical axis.

13						
12						
11						
10						
9						
8						
7						
6						
5						
4						
3						
2						
1						
	Green	Orange	Yellow	Blue	Brown	Red

4. a. Which color has the largest number of M&M's®?

b. Which color has the least?

5. Record the information for each group in the class below.

Group	1	2	3	4	5	6	7	8
Number of blue								
Total in bag								

a. Did each group have the same number of M&M's®?

b. What is the average number of blue M&M's® per bag?

c. What is the average number of M&M's® per bag?

6. Looking at the class summary, answer the following questions.
   a. Compare the percent for each of your group's colors with the class percentages.

   b. Do you believe that the M&M Company produces an equal quantity of each color. Explain.

   c. What other products can you think of which may have similar packaging plans that we could investigation?

7. The M&M Company reports that (as of 6-22-95) the production ratio for plain M&M's® is as follows:

Color	Percent
Green	10
Orange	10
Yellow	20
Blue	20
Brown	20
Red	20

a. Compare the color percentages in your bag to the company's standards.

b. Does the class percent of blue M&M's® fit the plan? Give evidence to support your answer.

8. Estimate the number of M&M's® in your instructor's bag.
Estimate the number of blue M&M's® in your instructor's bag.

Estimated total =	Estimated blue =
Actual total =	Actual blue =

9. You may now nibble!